KU-789-113

Nick Six

A Numberline Lane book
by
Fiona and Nick Reynolds

Nick Six was very excited.

That night he was going to play his guitar in a concert for all of his friends.

The concert was going to be held at The Add Pad, home of Gus Plus.

Nick Six had been practising for weeks and had learnt lots of new songs.

He had even written some songs himself such as "Count with Me!" and "Add on One."

Nick Six started to get ready.

He decided to have a bath, with plenty of bubbles in it.

He sang one of his favourite songs:

> "Count with me,
> One, Two Three,
> Just one more,
> Will make it four."

Just then Nick Six heard a noise.

It sounded like someone was playing his guitar – very badly.

He jumped out of his bath and ran downstairs.

There was his guitar lying on the floor.

The front door was wide open.

There was someone skipping down his garden path, but he couldn't see who it was.

It was Linus Minus.

He was chuckling to himself as he waved three guitar strings in the air.

Nick Six looked at his guitar again.

He counted the strings "One, two, three... Oh no!" said Nick Six.

"There should be six strings, but half of them are missing."

"I can't play my guitar with only three strings!"

Nick Six decided to visit Kevin Seven.

Kevin Seven could fix all sorts of things.

He picked up his guitar and walked along Numberline Lane until he had reached house Number Seven.

Kevin Seven was in his front garden, busy mending his gate.

"Hello Nick Six!" said Kevin Seven.

"Hello Kevin Seven," said Nick Six.

"What is the matter?" asked Kevin Seven.

"Linus Minus has taken away three strings from my guitar. I need my guitar tonight because I am playing in a concert at The Add Pad. I wondered if you could fix it for me?"

Kevin Seven scratched his head and had a long think.

He went into his shed and started looking through all his boxes of bits and pieces.

"I'm very sorry" said Kevin Seven. "I don't seem to have any guitar strings."

"We will have to go and visit Gus Plus. I'm sure he will be able to help!"

So off they went to The Add Pad.

Nick Six knocked on the door
"Knock–knock, knock–knock,
knock–knock".

The door opened and there stood Gus
Plus.

"Welcome, my friends, do come in. Is
there anything I can do to help?" he
asked.

Nick Six and Kevin Seven explained to Gus
Plus all about the guitar, and how Linus
Minus had taken away half of the strings.

"Don't worry," said Gus Plus. "I can help. All you need to do is double!"

"Double?" said Nick Six looking a little puzzled.

"You double a number by adding it to itself." said Gus Plus. "Let me show you with these cakes."

Gus Plus went to the table where there were two cakes on a plate.

He took out his number wand and said some magic words.

With a flash of light and a puff of smoke two more cakes appeared on the plate.

"You see, two add two is four - there are double the number of cakes now!"

"If you double the number of strings on your guitar you will have six strings again, because three add three makes six," said Gus Plus.

Nick Six and Kevin Seven watched as Gus Plus waved his magic wand.

With a flash of light and a puff of smoke three new strings appeared on Nick Six's guitar.

"Thank you very much" said Nick Six. "Now I will be able to play my songs."

All the Numbers from Numberline Lane came to the concert.

Nick Six sang his songs and played his guitar, and everyone clapped and cheered!